The Vaquita

The Biology of an Endangered Porpoise

Written by Aidan Bodeo-Lomicky

Illustrated by Aidan, Cooper, and Skylar Bodeo-Lomicky

For Dr. Thomas Jefferson, for answering my first and every subsequent Vaquita question,

And for my family and friends, for accepting, dealing with, and embracing my obsession.

Copyright © 2013 Aidan Bodeo-Lomicky

www.vlogvaquita.com

All rights reserved.

First Edition
Published June 2013

Cover illustration copyright © 2013 Guillermo Munro Colosio, aka Memuco

Back cover photograph copyright © 2008 Dr. Thomas A. Jefferson, PhD

Book illustrations copyright © 2013 Aidan, Cooper, and Skylar Bodeo-Lomicky

No part of this book may be reproduced or used for any purpose or by anyone without permission, the only exceptions being for small excerpts used for publicity or marketing.

This book provides information about an endangered porpoise.

Table of Contents

What is the Vaquita?..5

Appearance...6

Distribution..8

Taxonomy..11

Food and Feeding Habits...14

Behavior..16

Reproduction...18

Threats..20

    Gillnets..20

    Other Threats..23

Poems...24

    Conquer the Net..24

    Vaquita – The Life of an Endangered Porpoise................25

Suggested Reading...28

Acknowledgements...29

About the Author...30

A surfacing Vaquita

What is the Vaquita?

The Vaquita, *Phocoena sinus*, is the smallest species of cetacean, or whales, dolphins, and porpoises. It is dark grey on its back and sides, fading to light grey or white on its belly. With dark patches over its eyes and mouth, the Vaquita almost looks as if it is wearing Goth makeup. It is also the rarest and most endangered marine mammal, with only around 200 left in the world! It has an extremely limited distribution, only living in the northernmost tip of the Gulf of California, also known as the Sea of Cortés. In 1958, the Vaquita was recorded based on a few skulls, but no live Vaquita had been seen or fully described until 1985. It is Mexico's national marine mammal. The Vaquita is so rare because it often gets tangled in gillnets set out for fish and shrimp, is trapped below the surface, and dies. Several Vaquitas die this way every year! The Vaquita could go extinct in the very near future if we don't help it! In this book you will learn all about these little-known porpoises, and things you can do to help save them.

Appearance

The Vaquita has the largest dorsal fin (which is shark-like and about 7 inches tall) and flippers in proportion to its body length out of all the porpoises. Adult Vaquita can weigh up to 120 pounds, and are quite robust. An average male is 4.6 feet long, while the female is usually around 5 feet long. Depending on the light conditions, a Vaquita can appear olive or tawny brown, and many observers just describe them as "dark." But in adults, the true color of the dorsal surface is dark grey, the sides fading to pale grey, with a white ventral surface (belly). This porpoise has a large dark ring around each eye and a like-colored patch on the lips. There is a thin grey stripe from the edge of the mouth patch to the flippers. The overall coloration is darker in newborns than in adults, particularly in the head and in the areas behind the eyes. A Juvenile has whitish spots on the leading edge of its dorsal fin, which eventually turn into bumps as the Vaquita ages. When seen at a distance, the tall dorsal fin of the Vaquita can cause confusion with that of the Bottlenose and Common Dolphins, because they can both be found in the Vaquita's range. But the small pod and body size, lack of prominent beak, and behavior should usually allow the Vaquita to be easily identified. The Vaquita has a subtle yet intricate and beautiful color pattern, and its head has a rounded appearance, with no obvious beak or melon. The mouth patch creates a permanent "smile," giving it a deceivingly happy appearance.

A typical adult Vaquita

Distribution

The Vaquita has the most limited range of any marine cetacean, though its habitat appears to be very clean and healthy. The Vaquita is one of only two porpoise species that live in warm waters. The Vaquita is unique in that it can tolerate large temperature fluctuations (which could be the reason for its seasonal reproduction). The Vaquita lives in only one place on Earth: in the shallow waters of the northernmost tip of the Gulf of California (Sea of Cortés), in Baja California, Mexico, where the bottom sediment is clay and silt. This tiny distribution area is only about 1,400 square miles (4,000 sq. kilometers), ¼ the size of metropolitan Los Angeles, and is known as the PACE-Vaquita region. In this minute range, a dangerously small number of Vaquita actually live—around 200 individuals. There are no other Vaquita left in the world because none has proven to survive in captivity.

In part to help protect the Vaquita, the Upper Gulf of California Biosphere Reserve was created in its core range in 1993, and an additional Vaquita Refuge was created in 2005. This makes it illegal to use gillnets to fish there. While this is good news, the two refuges combined still do not cover the entire Vaquita range. Furthermore, since the fishing rules are not strictly enforced, many fishermen are still using gillnets there. It is therefore vital that the Mexican government strictly enforces the gillnet ban, (espe-

Range of the Vaquita

cially large mesh nets), and also completely bans gillnet fishing in the Vaquita's *entire* range, not just a portion of it. The Vaquita is most commonly found in the shallow murky water south of the Colorado River delta, about a 4-hour drive from San Diego, with the highest densities occurring around Rocas Consag, which is off the coasts of the Mexican cities of San Felipe, Puerto Peñasco, and El Golfo de Santa Clara. It is believed that the Vaquita used to live in an area farther south along the Mexican mainland. The land on both sides of the Gulf is covered in hot sand, giving the Vaquita the nickname "Desert Porpoise." Some other names for the Vaquita include: Panda of the Sea, Gulf of California Harbor Porpoise, Gulf of California Porpoise, or Gulf Porpoise (English), *Cochito* or *Vaquita Marina* (Spanish), *Pazifischer Hafenschweinswal* (German), and *Marsouin du Golfe de Californie* (French). Vaquita is Spanish for "Little Cow." The Vaquita is Mexico's only endemic marine mammal. It is extremely important that we keep the Vaquita's habitat safe, because none has ever survived in captivity.

Taxonomy

The Vaquita was first described in 1958 by Ken Norris and William McFarland. It has no subspecies, likely because of its limited distribution. The scientific name of the Vaquita is *Phocoena sinus*, which loosely translates to "Gulf pig-fish" in Latin. The Vaquita is not known to migrate. The Vaquita is considered an EDGE Species, which means Evolutionarily Distinct, Globally Endangered. Evolutionarily Distinct means it represents more of the tree of life than the average animal. There are 7 species of porpoise in the world, in 3 genera: the Vaquita, the Narrow-ridged Finless Porpoise, the Indo-Pacific Finless Porpoise, the Harbor Porpoise, the Burmeister's Porpoise, the Spectacled Porpoise, and the Dall's Porpoise. The one most closely related to the Vaquita is the Burmeister's Porpoise (which lives close to the shores of much of South America), although they look quite different from one another. The Burmeister's Porpoise is dark brown-grey to black with a backwards-leaning dorsal fin, while the Vaquita is lighter grey with a tall, falcate dorsal fin. Porpoises are very similar to dolphins, but are generally smaller and more shy (porpoises do not perform the acrobatic stunts that dolphins are famous for), with spade-shaped teeth, rather than the dolphin's conical ones. Porpoises have no prominent beak, unlike most dolphins. You need good eyes and a lot of luck to see a wild porpoise!

Vaquita

Narrow-ridged Finless Porpoise

Indo-Pacific Finless Porpoise

Harbor Porpoise

Burmeister's Porpoise

Spectacled Porpoise

Dall's Porpoise

The Members of the Porpoise Family
All illustrations drawn to scale

Food and Feeding Habits

Like other porpoises, the Vaquita is a very shy, relatively slow marine mammal. It is near the top of the food chain in the Gulf, but Great White, Mako, and Blacktip Sharks, among other shark species, have been found with Vaquita parts inside their stomachs. Some Vaquita that were found tangled in gillnets showed scars on their flukes from teeth that could have belonged to a shark or Killer Whale (Orca), however there have been no reports of direct attacks on Vaquita by these species of sharks or Killer Whales. The Vaquita, like many other cetaceans, uses sonar pings, called echolocation to find food. However, even this incredible survival method cannot reliably sense the gillnets that are so infamous for killing Vaquita. It feeds in a leisurely manner, almost always avoiding being seen. They have 66 to 84 teeth, with 32-44 teeth in the top jaw, and 34-40 in the bottom. From contents found in Vaquitas' stomachs, it is known that it eats a large variety of local fish such as Gulf Croakers and Bronze-striped Grunts. It has also been known to eat small crabs and squid. It is nearly always seen in shallow waters (less than 100 feet deep, and within 16 miles of shore). It was previously thought that the damming of the Colorado River was restricting the nutrients (and therefore food) in the Vaquita's ecosystem. This theory, however, was disproven when scientists found an abundance of zooplankton and like creatures in the Gulf's waters. Still, relatively little information has been gathered about the food and feeding habits of these very elusive creatures.

The Vaquita's place in the food chain

Behavior

Because one has to be both extremely lucky *and* observant to get even a glimpse of a live Vaquita, there is not that much information known about the behavior of the Desert Porpoise. We do know that it navigates the Gulf and communicates using sonar, and is not known to breach. It breathes every minute or so. To breathe, a Vaquita rises with a slow, forward-rolling movement that barely disturbs the water's surface, and then disappears quickly, often for a long time. It has an indistinct blow, but makes a loud, sharp puffing sound similar to that of a Harbor Porpoise (which is geographically the closest porpoise to the Vaquita, along with being the most similar in appearance). It does everything in its power to avoid boats, but when a Vaquita *is* seen, it is usually only once for just a few seconds. Vaquita are usually spotted in small groups of one to three individuals. Also, it is rare to see more than just the back and dorsal fin of the Vaquita. Most often, the pod is comprised of a mother (females are called cows) and her calf, and occasionally some other relatives. When larger groups are seen, it is usually the result of an aggregation formed by many small subgroups spread out over a few thousand square meters. The oldest known Vaquita lived to a ripe old age of 21 years.

A Vaquita in the calm

Reproduction

Every *other* year, in March or April, after an 11-month gestation, a precious, 2-foot-long Vaquita calf is born. The tails emerges first, because if the head came out first the calf would drown in the time it takes for it to completely emerge. Its mother helps it breathe by nudging it up to the surface to take its first inhalation of fresh sea air. Porpoises are born with lines called fetal folds that are caused by the calf's position in the womb, which disappear soon after birth. The calf is nursed by the mother for 6 months to a year. Then it is finally weaned off milk and joins its mother's pod. After about 6 years, it reaches sexual maturity. It then mates in the spring or summer. The average lifespan of the Vaquita is around 20 years, meaning a female could have up to 7 calves in her lifetime! Vaquita do not mate for life (no cetaceans do). During the breeding season, a male (bull) Vaquita attempts to mate with as many females as it can, increasing the chance of a successful birth with that male's genes. Virtually nothing is known about most of the Vaquita's life, so a lot of information is based upon knowledge of other porpoises. But what we do know is that the Vaquita is disappearing fast.

Mother and Calf

Threats

The Vaquita used to be the second-most endangered cetacean, next to a river dolphin called the Baiji. With the recent extinction of the Baiji, the Critically Endangered Vaquita received the dubious title of most endangered cetacean, with only around 200 individuals remaining. And that number is quickly decreasing. It is quite clear that there are only a few threats facing the Vaquita. This chapter is divided into two parts: Gillnets and Other Threats.

Gillnets

Gillnets are undoubtedly the main cause of Vaquita mortality. In the past, they were responsible for 49 to as many as 84 Vaquita deaths per year, which has surely helped drive the species toward extinction. There are many kinds of gillnets, but all are dangerous to the Vaquita. Gillnets are nearly invisible nets used to catch fish and shrimp. Previously, they were used to catch the fish, Totoaba, a large member of the drum family. Today, the Totoaba is nearly extinct as well, due to overfishing. During the peak of Totoaba fishing is when the Vaquita's decline really accelerated. Gillnets are the most efficient and cheapest fishing method available to the fishermen of the Gulf, so one can easily see the essence of the problem. Vaquita either swim into the nets accidentally, or see the fish or shrimp already in the net and try to catch one, resulting in the same tragedy. Once a Vaquita makes contact with the net, its natural reaction is to roll, which only makes things worse. So once a Vaquita gets caught in a net, it's not coming out alive. The obvious and only way to save the Vaquita is to get gillnets out of the water. Some efforts have been made by the Mexican government in achieving this goal. Aside from the creation of the Vaquita Refuge and Biosphere Reserve, a program was created called the buy-out. The buy-out program consists of the buy-out, the rent-out, and the switch-out. The buy-out is when the fishermen are paid to turn in their fishing permits, but they keep their boats and nets, so they can still illegally fish if they are so inclined. The rent-out is when the fishermen are paid to stay out of the Refuge (though they keep their permits). The most effective and hopeful option is the switch-out, in which they are given an alternative fishing method that is Vaquita-safe, such as pots, hook and line, long line, and small trawls. Small trawls have recently been developed and tested specifically for the Vaquita, and they are proven to be just as effective as gillnets, but are Vaquita-safe. This is a major breakthrough in Vaquita science, and in June 2013, the Mexican government announced that they are planning to completely switch out all gillnets within the next 3 years. Another budding idea in the Gulf is ecotourism. Ecotourism is a form of tourism in pristine environments used as an alternative to mass tourism. Its main focus is to educate the traveler and/or raise funds for the area. Ecotourism would be an ideal career switch for fishermen, the people who know the Gulf best. All of these things are new ideas and will take more time to develop. Unfortunately, time is one luxury the Vaquita does not have. Money may be the only thing that can accelerate the progress of turning these ideas into reality.

Dead Vaquita in a gillnet

Other Threats

Besides gillnets, there are no definitive threats facing the Vaquita. However, there have been some suggested threats that could possibly affect the Vaquita population. The main one of these is the lack of fresh water flow to the Gulf due to the damming of the Colorado River Delta in the early 1900's. Upon recent study, however, it is thought that the Pacific Ocean provides a perfectly healthy level of nutrients to the Gulf, and the Colorado River is not really necessary for the health of the Vaquita and its habitat. Also, inbreeding depression and environmental contamination have a possibility of affecting the Vaquita population. Though all of these things are or were thought to be threats to the Vaquita, gillnets are much more dangerous, so it is best not to focus any time or attention on these minor threats. The Vaquita population could be declining at 7% or more a year, and this is solely due to gillnet fishing. If Vaquita numbers drop below 100, the species' survival is very unlikely. Another conservation effort is the creation of CIRVA, a group that recommends more ideas to save the Panda of the Sea. In the eyes of CIRVA, the Vaquita can be saved. This book's profits will go directly towards Vaquita research and education.

Conquer the Net

By Aidan Bodeo-Lomicky

Through thick rustling leaves of beige and toast,
O'er crisp vast ice whiter than a ghost.
Down streetways and alleys swarming with crowds,
Up huge frosty mountains piercing the clouds.
Down rift valleys and 'cross frozen tundra,
African deserts and the Land Down Under.
The world is huge, with room to spare,
But something's somewhere, and only there.
Dive in the sea, sink like a fallen ship.
Swim until you reach the southernmost tip
Of California, then head through the foam,
And find the place Vaquita call home:
The Sea of Cortés, rich and warm,
With rainbow fish teeming in swarms.
The tiny Vaquita, gentle and few,
Are vanishing quickly; what do we do?
They happily swim 'mong coral and kelp,
In spite of this, they need our help.
Gillnets trap them and take their lives,
Until now, we've ignored their strife.
Be brave, la Vaquita, and do not fret.
Side by side, we'll conquer the net.

Vaquita – The Life of an Endangered Porpoise

By Aidan Bodeo-Lomicky

BIRTH

Cries of gulls pierce the air,

The Gulf of Cali has beauty to spare.

The April sun warms the water,

The Sea of Cortes could not be hotter.

Although the sky and land are great,

What's below the surf is worth the wait.

Plunge in the ocean, go down a few feet.

A submerged birth scene is what you will meet.

There, in the seaweed, floats a mother,

Giving birth to her young calf's brother:

The tail comes first,

Then out comes the head.

The calf is free,

But sinks like lead;

The mother is nervous.

But all is well,

The calf's not ill,

Things are swell.

To have healthy kids, some would kill.

The mother tends to her precious boy,

Tiny and cute, he resembles a toy.

The calf takes in the blazing sun,

With no idea of what's to come….

ADOLESCENCE:

A few months now have already passed,

The calf's older, he's not a tot, at last!

He's finally stopped cowering under his mom.

Now he's swimming around with aplomb.

He learned to catch fish all on his own,

He learned how to squeak, whistle, and moan.

He can log, spyhop, breathe, and breach,

Echo-locate things his eyes can't reach.

His days of being cared for are fading away.

He will join his mother's pod any day.

The clock is ticking fast, and in essence,

These are his last days of adolescence.

MATURITY

Bubbles stream down his graceful form.

The sea surrounds him, rich and warm.

He playfully darts through the silky water,

He finds a mate and they have a daughter.

They all live together with his brother and mom.

The pod's noise meets that of a bomb.

Their lives are all perfect, they whistle nice songs.

Nothing. Yes, nothing could ever go wrong.

DEATH

As you know, every great thing must end.

Do not dread death, just savor time you spend

With family and friends, and even foes.

Life is more powerful than anyone knows,

But death ends life much faster than a cheetah.

How does this relate? Let's meet our Vaquita.

He loves his life, frolicking with loved ones.

He lives with parents, daughters, and their sons.

La Vaquita is living his life at its best,

Everything's perfect, this Vaquita is blessed.

Think about something while he thinks about fish:

There must be a reason this is called what it is.

Ok, fine, I'll go back to the story.

While the pod is living in glory,

They dive and blow and breach and float.

But just 'round the corner approaches a boat.

Our little Vaquita sees a school of fish,

He dashes towards the boat, that he didn't I wish.

Suddenly, out of nowhere, he gets caught in a net,

He rolls frantically and starts to fret.

His heart stops.

He's gone.

But do not dread death, here's what some say:

Life after death is just another day.

This may be true for some forms of life,

But not so with the Vaquita's strife.

The Gulf without the Vaquita is like a forest without moss.

When one Vaquita dies, the entire species may be lost.

Suggested Reading

- **Marine Mammals of the World: A Comprehensive Guide to their Identification –** Thomas A. Jefferson, Marc A. Webber, Robert L. Pitman
- **¡Viva Vaquita!** – www.vivavaquita.org
- **Save the Whales** – www.savethewhales.org
- **V-log** – www.vlogvaquita.com
- **Muskwa Club –** www.muskwaclub.org

To learn how to help the Vaquita, please visit all of these websites and read the field guide. I used these sites (or links found on the sites) and guide to get all of the information for this book. V-log is my blog that I created as a platform to share information, art, writing, and news about the Vaquita.

All profits of this book will go to ¡Viva Vaquita!, the leading Vaquita conservation group, and other Vaquita-related organizations, education, research, and conservation.

Acknowledgements

This book was made possible by a few amazing individuals. Thank you to these people and to anyone who has ever helped the Vaquita.

First, I would like to thank Dr. Thomas A. Jefferson, PhD, for providing the back cover image, much of the information, editing the book's content, and a whole lot of inspiration and assistance. He answered my first ever Vaquita question and hasn't stopped since. He has participated in endless Vaquita expeditions, talks, booths, and much more. When I grow up, I want his job!

Second, I would like to acknowledge my family, for helping and supporting me throughout the writing and production process: my dad for always being my grammatical editor; my mom for helping me get the illustrations and author photograph onto these pages and making them look good, along with helping me with the publishing process; my younger sister Skylar for being one of the biggest Vaquita supporters and animal lovers I know (she also drew the pictures next to the suggested reading, acknowledgements, and the "Conquer the Net" poem); and my younger brother, Cooper, for doing many of the illustrations. He is a talented up-and-coming artist who has dedicated many hours to this book.

Also, I would love to thank the extremely awesome Mexican artist Guillermo Munro Colosio (Memuco) for voluntarily illustrating this book's cover. He is such a great artist who has a deep passion for the natural world.

Finally, I would like to acknowledge William Whittenbury and his mom Beth for everything they have done. William is the president of the Muskwa Club, a student-run group that I am now part of, and has helped the Vaquita so much. Beth introduced me to my self-publisher, which pushed me to make this book a reality.

About the Author

Aidan Bodeo-Lomicky, born in 2000, lives in Bethlehem, Pennsylvania. He plays tennis and trains almost 20 hours a week, and aspires to become professional tennis player one day. He also does a lot of birding, and loves being outdoors. But perhaps his favorite thing to do is Vaquita conservation.

Aidan has been working on Vaquita conservation since 2010. He runs a blog called V-log, http://vlogvaquita.com/, on which you can find all of the information you need about the Vaquita, along with everything that he is currently working on.

Aidan has many ideas in the works, and has a lot of associates and supporters who are helping to make his ideas reality(some of whom are listed in the acknowledgements). His goal is to save the Vaquita while allowing the Gulf of California fishermen to continue to make a living.

Made in the USA
Charleston, SC
01 July 2013